Discovering Cities

Manchester

Photo: University of Manchester

Metrolink.
Photo: University of Manchester

G-Mex Centre.
Photo: G-Mex Ltd.

Public art outside the new Marks & Spencer.
Photo: University of Manchester

The Lowry.
Photo: Len Grant/The Lowry

The view north-east from Beehive Mill.
Photo: Aidan O'Rourke

The Opera House.
Photo: Aidan O'Rourke

Corporation Street.
Photo: Aidan O'Rourke

Discovering Cities

Manchester

Christopher M. Law

Series Editors
Peter S. Fox and
Christopher M. Law

Preface

© Len Grant/The Lowry

The variety and complexity of cities as revealed in their built form has been a source of fascination to the local resident and visitor alike. Can a clear spatial structure be discerned? Why do activities cluster in distinctive quarters or zones? How do relict features throw light on the constantly evolving city?

For a long time, human geographers, regional economists, urban sociologists and local historians have sought to understand the processes which shape the city. The growth (or decline) of the city is affected by local, regional and global economic forces. The forces which shape the internal structure of the city are many and varied. There is a market in land which influences the pattern of land use and change. Public policies are often significant but can be complex to understand and difficult to follow. Social factors such as those of class and ethnic community identities are also important.

Written by urban geographers with vast knowledge and experience of the city in question, *Discovering Cities* gathers these issues together in concise and practical guides, illustrated with colour maps and photographs, to enable an enhanced perspective of cities of the British Isles.

Peter S. Fox, Chilwell Comprehensive School, Chilwell, Nottingham

Christopher M. Law, Division of Geography, University of Salford, Salford, Manchester

Acknowledgements

The author would like to thank Gustav Dobrzynski for drawing the maps, and Paul Hindle for reading the first draft.

ISBN 1 899085 97 1
First published 2001
Impression number 10 9 8 7 6 5 4 3 2 1
Year 2003 2002 2001

Published by the Geographical Association, 160 Solly Street, Sheffield S1 4BF.
Website: www.geography.org.uk
E-mail: ga@geography.org.uk

The Geographical Association is a registered charity: no 313129.

The Publications Officer of the GA would be happy to hear from other potential authors who have ideas for geography books. You may contact the Officer via the GA at the address above. The views expressed in this publication are those of the author and do not necessarily represent those of the Geographical Association.

Editing: Rose Pipes
Design and typesetting: ATG Design, Leeds
Cartography: Gustav Dobrzynski
Printing and binding: Colorcraft Ltd, Hong Kong

Contents

© University of Manchester

© John Peters

© Aidan O'Rourke

Introduction

Manchester is one of Britain's great cities. Long known throughout the world because of its pioneer role in the Industrial Revolution, the city is a major regional centre, at the heart of a conurbation with a population of 2.5 million.

According to Asa Briggs in his book *Victorian Cities* (1968), in the first half of the nineteenth century Manchester was the 'shock city' of a new era and the 'symbol of the new age' with distinctive social relations between entrepreneurs and a new urban working class. In the nineteenth century its fame as the archetypal industrial city was enhanced through the writings of authors such as Friedrich Engels in *The Condition of the Working Classes in England* (1845) and Elizabeth Gaskell, with her novels *Mary Barton* (1848) and *North and South* (1855). This industrial image has lingered on through the bleak landscapes of the artist L.S. Lowry, peopled by matchstick figures, and the television soap *Coronation Street.* Currently, Manchester is doing its best to shake off this image and to reinvent itself as a dynamic, modern, enterprising and truly European, if not international, city. While best known by many as the home of Manchester United Football Club, or for its club and pop music scene, the city is also a major centre of industry, science, higher education and cultural activities, and has the largest airport outside the London region.

What is Manchester?

Administratively, the City of Manchester is a local government unit which stretches north and south of the city centre from Blackley to Wythenshawe to form an elongated area with a population of about 430,000 (1998 estimate). It is hemmed in by other administrative units, notably by the City of Salford on its west side (Figure 1). The boundary between these two cities, mostly along the River Irwell, dates back a thousand years to a time when Manchester was a mere parish in the Hundred of Salford. At Victoria Bridge, inner city Salford adjoins Manchester's city centre. Although Manchester is now the more important of the two cities, there has been intense rivalry between them in the past which still lingers today. The built-up area around Manchester, the conurbation, includes many independent and formerly discrete towns such as Bolton, Oldham and Stockport which have become joined up in the urban sprawl of the twentieth century. The administrative reforms of 1974 abolished many small units and created ten metropolitan districts focused (with the exception of Trafford and Tameside) on the larger towns. Many of the communities which are outside the conurbation, but within a 24km radius, have very close relationships with the core area, notably via commuting, and these comprise a wider Manchester region.

Figure 1: The administrative units of Greater Manchester County.

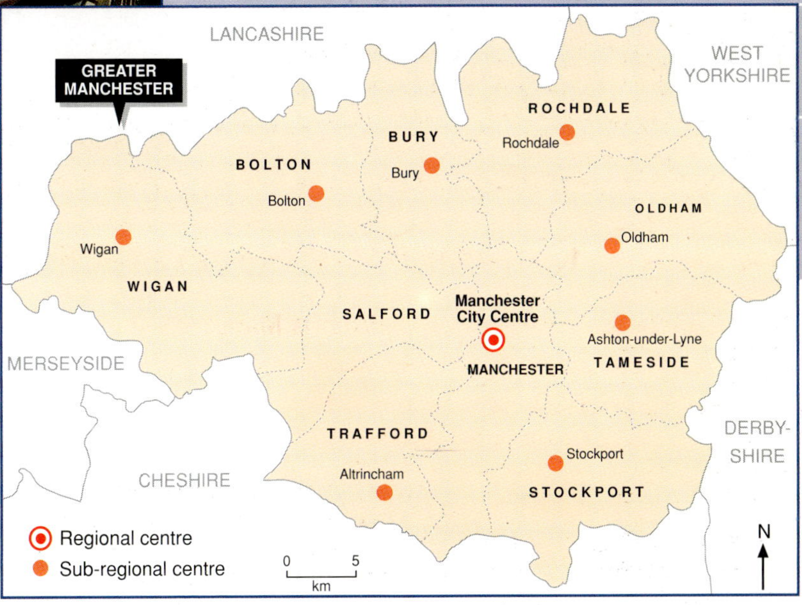

The development of Manchester

Manchester's growth as a major industrial centre took place during the 250 years following the Industrial Revolution It is this period that shaped most of the physical development that can be seen today, and this industrial phase that left many problems which remain unresolved.

The built environment is constantly being reworked through the processes of capitalism as applied in particular to property development, and through public policies which seek to control development in the community's interest. There is a great deal that is of interest to the geographer in these processes and their consequences, but I will concentrate on the city centre and inner city.

Historical geography

Contents

© Aidan O'Rourke

© University of Manchester

© Aidan O'Rourke

© University of Manchester

Figure 2: The situation of Manchester.

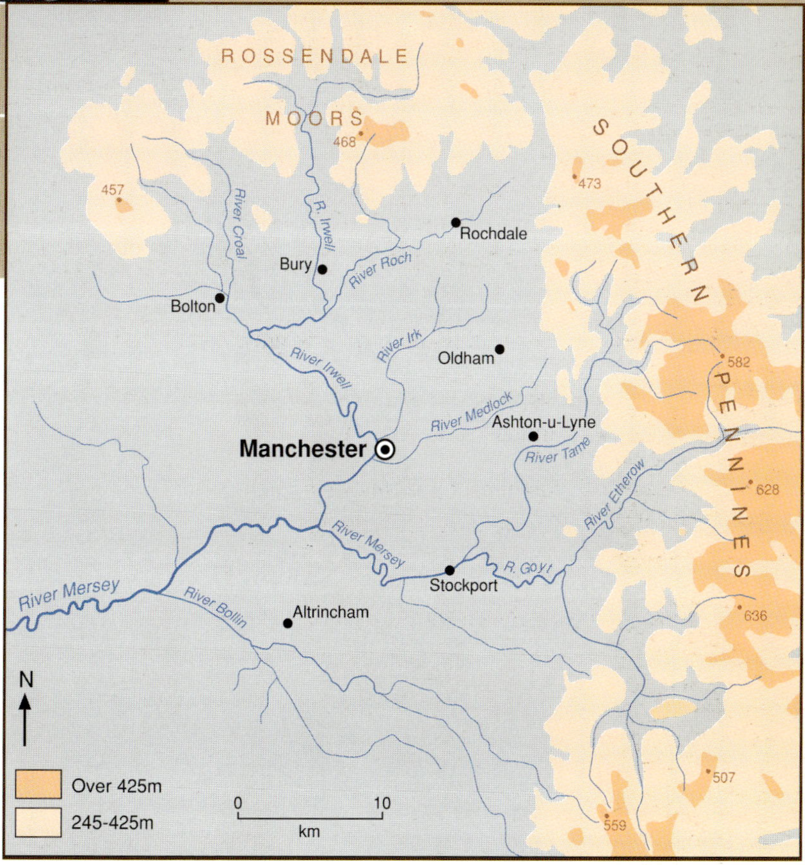

Site, situation and early growth

Manchester lies in an embayment on the west side of the Pennines. Streams flow from the Rossendale hills to the north and the Pennines to the east and join to form the River Mersey which runs westwards to its estuary at Liverpool about 60km (Figure 2). The principal tributary of the Mersey is the Irwell which flows southwards from Bury to Manchester. Small market towns including Bolton, Bury, Rochdale and Stockport developed where streams left the hills. At the centre of this area Manchester developed as the principal market town.

The Romans built the first known settlement (Mamucium) in the area on firm, rising ground above the confluence of the Rivers Medlock and Irwell (Figure 3). The remains of Mamucium became Castlefield. Medieval Manchester developed higher up the Irwell to the north, again at the confluence of two rivers, the Irk and Irwell (close to Victoria Station and the Cathedral). A bridge (later named Victoria) linked the small settlement to Salford on the other side of the Irwell, and a market place (about where Marks & Spencer now stands) evolved south of the parish church, which was later to become the Cathedral. This building, and Chetham's College to the immediate north, are the only surviving medieval buildings.

Figure 3: The sites of Roman and medieval Manchester. Adapted from Frangopulo, 1962.

The general situation of this ancient settlement does not immediately suggest nodality in terms of national routes. Although the Roman settlement lay on a cross-Pennine route, it was well to the east of the main north-south route which passes through Warrington. In more recent times Manchester's nodality was improved by the construction of the Mersey and Irwell navigation in 1734, turnpike roads, canals (from 1764 onwards), railways (from 1830 onwards), the airport and motorways.

Manchester and the Industrial Revolution

The transformation of Manchester from a small market town to a leading industrial city and major regional centre began in the late eighteenth century in the period known as the Industrial Revolution. Directly linked to the rise of the cotton textile industry, Manchester became the leading manufacturing and commercial centre in England – hence its nickname of 'Cottonopolis'. The commercial activities relating to the industry were centred on the Royal Exchange and the zone of warehouses. While spinning mills developed from the 1770s, power-loom weaving did not become general until the late 1820s. In this early period yarn was bought by Manchester merchants, stored in their warehouses and sold on to hand-loom weavers, bought back as cloth and sold on to clothiers. Even after the development of

Table 1: Statistics of the UK Cotton Industry. Source: Mitchell, 1988.

Year	Raw cotton imports (m lbs)	Raw cotton consumption (m lbs)	Cotton cloth exports (m sq yards)
1775	7	–	–
1800	56	52	–
1819	151	109	203
1913	2174	2178	7075
1939	1434	1317	1393
1951	1166	1024	900
1965	614	506	207
1980	276	152	172

power-loom weaving the system of Manchester merchants and their warehouses was to remain important in the industry. The zone of warehouses around the office and retail core was, and still is, a distinctive part of the spatial structure of the city centre.

The Industrial Revolution in the north-west of England was supported by, and in turn stimulated, new transport developments. River navigation was improved and turnpike roads were developed. One of the first canal developments in the region was the Duke of Bridgewater's canal from Worsley (about 13km west of Manchester) into the city at Castlefield, completed in 1765. In 1800 it was linked to the Rochdale Canal which ran between Manchester and Sowerby Bridge on the Yorkshire river network. On the north-eastern edge of the city centre (near the present Piccadilly Station) the Ashton Canal was built and later linked to the Huddersfield and the Macclesfield Canals. Following a period of disuse, the Huddersfield and Rochdale Canals are now being restored.

The first successful steam passenger railway in Britain was the Manchester to Liverpool line, opened in 1830. Soon Manchester was linked to most towns in the region and, via Crewe, to the main west coast line linking London to Scotland. In the nineteenth century each major railway company sought to have its own terminus in the city with the result that four stations were built: Piccadilly, Victoria, Exchange and Central. None of these penetrated the heart of the city centre and, despite many proposals, this problem was not solved until 1992 when Metrolink was opened, running trams from the stations through the city centre.

The cotton industry grew rapidly to become one of the UK's major economic activities (Table 1). By the mid-nineteenth century cotton constituted 44% of all UK exports. Two consequences of its expansion were population growth and urbanisation. The population of Manchester (and Salford) grew rapidly from the late eighteenth century. Estimated at 20,000 in 1756, it reached 388,000 by 1851. The population of the cotton towns in the wider region was also growing.

Manchester at the beginning of the twentieth century

The cotton industry reached its peak just before the First World War, although its rate of growth had slowed down considerably by then (see Table 1). About 620,000 people were employed directly in the industry within the region, and the number indirectly dependent on the industry, including workers in chemicals, engineering, coal mining,

© Aidan O'Rourke

transport and trading companies as well as the service trades supplying the cotton workers was considerably larger. The built-up area around Manchester extended to a radius of 5km and had a population of about 1 million. This did not include the detached suburbs, such as Sale, Altrincham, Cheadle and Wilmslow. Compared to the north of the city, these southern suburbs and their surrounding areas were largely devoid of industry, were dryer, and largely avoided the industrial pollution as the predominant winds blow from the south-west. The area within a radius of 16-19km from the city centre had a population of about 2.3 million and the population of the area within a 90km radius had a population of nearly 10 million. It included West Yorkshire, parts of the East Midlands, the Potteries and north-east Wales, as well as the rest of north-west England. The population was higher in this area than in an area of similar size around London. This centrality to northern industrial England enabled Manchester to play an important role in the economy of the region.

Manchester attained its greatest relative importance between 1890 and 1910. It was the undisputed capital of an industry and region which played a very important role in the life of the country. Its political influence was such that the city could claim that what Manchester does today the rest of the country will do tomorrow. Among the major developments that took place was the opening of the Manchester Ship Canal. This provided a link between the docks (actually located in Salford and Old Trafford) and the Mersey estuary. Two years later, in 1896, an innovative industrial estate, the first in the country, was opened beside the Ship Canal on land belonging to the former private estate of Trafford Park. A grand new Town Hall, designed by Waterhouse, was opened in 1877, and the art gallery in 1882. Its orchestra, the Hallé, was the first in the country.

Unlike towns in the region whose economies were dominated by cotton, Manchester (and Salford) had a diversified economic structure. Nevertheless, the cotton industry had a dominant influence: there were cotton mills in various parts of the built-up area including Ancoats just north of the city centre, and much of the commercial activity in the city centre was related to cotton. Trading was organised around the Royal Exchange and there was a ring of warehouses within a kilometre of the city centre, close (in an age of horse and carts) to the railway goods yards. As the cotton firms of the region merged or expanded through takeovers they required landmark head office buildings in Manchester, of which some remain. King Street was the location for the headquarters of many regional banks and for a Bank of England branch.

Figure 4: Models of the spatial structure of Manchester in the nineteenth century. Reproduced with permission from Dennis, 1984.

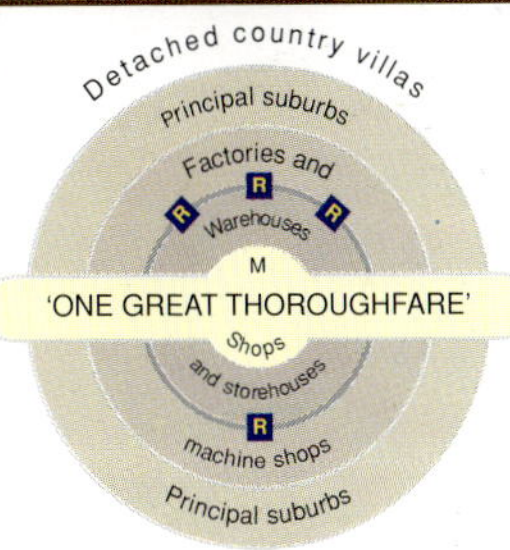

FAUCHER (1844)

R Railway terminus

M Municipal buildings

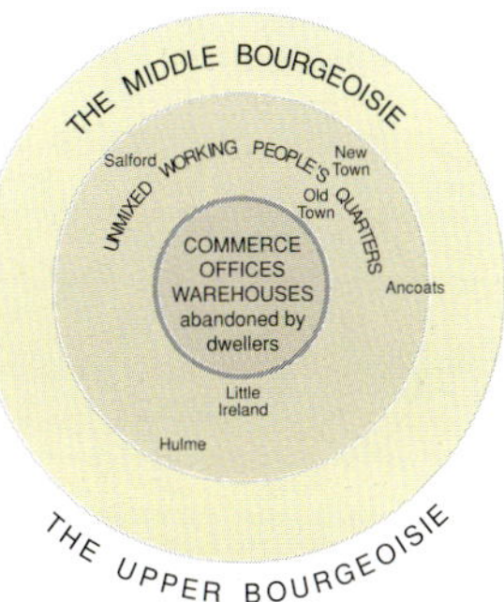

ENGELS (1844)

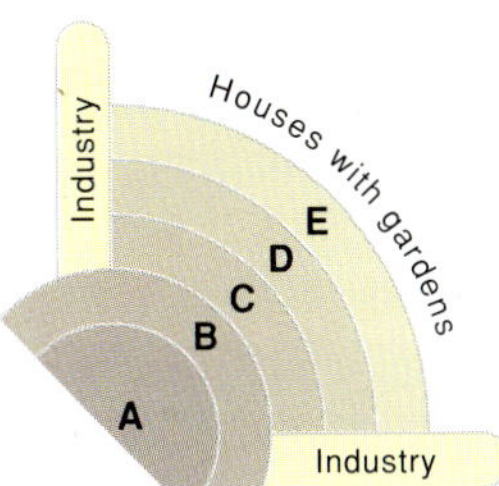

MARR (1904)

A Preindustrial core.

B C18th houses, now offices, workshops, lodging houses.

C Early C19th suburbs, now slums, factories and C18th villas.

D Bye-law housing, sanitary but on the downgrade.

E Better working-class housing.

Associated with the cotton industry was clothing manufacture, particularly women's clothes and waterproof coats. This industry was found in the northern part of the city centre and in adjacent parts of Manchester and Salford. Outside the city centre there were important industrial zones. To the east (Gorton, Openshaw and Beswick) were heavy industries such as iron and steel, railway locomotive manufacture and engineering, while to the west, along the river and in Trafford Park, was a zone ranging from heavy engineering to food (including five flour mills beside the canal).

The social spatial structure of Manchester

By the 1830s, the wealthier citizens of Manchester were moving out of the city centre. The houses in Mosley Street and King Street were being redeveloped for warehouses or banks. A new pattern of land use was evolving and foreign observers such as Engels, Marr and Faucher were to describe its characteristics (Figure 4). Engels observed that the rich lived on the outskirts of the city and that when they travelled to the centre they passed through the inner working-class suburbs which were often concealed from view by the strip pattern of shops

Table 2: Greater Manchester employment change. Source: Greater Manchester Research Unit (using Census of Employment).
NB. Figures may not add up due to rounding.
**No comparable figures because of changes in classification.*

1 Total employees

	All	Male	Female
1965	1,298,000	804,000	493,000
1977	1,150,000	667,000	484,000
1984	995,000	545,000	450,000
1987	1,033,000	556,000	477,000
1995	982,000	492,000	490,000

2 Industrial structure

	1977	1984	1995
Agriculture, etc	3000	2000	2000
Energy and water	24,000	21,000	7000
Manufacturing	446,000	302,000	205,000
Construction	62,000	50,000	39,000
Distribution, hotels, catering	*	200,000	214,000
Transport and communications	67,000	59,000	65,000
Banking, finance, business services	53,000	81,000	162,000
Other services	*	280,000	288,000
Total	1,150,000	995,000	982,000

along the main roads. Much of nineteenth century Manchester conformed to this concentric pattern of land use, the one exception being the wealthy suburbs of Upper Broughton and Prestwich on the ridge to the north followed by the Bury New Road.

Manchester in the middle of the twentieth century

The decline of the cotton textile industry after the First World War created enormous problems for the region and the city, many of which remain unresolved. Adaptation has been slow, partly due to insufficient input from government grants. Among the reasons given for this lack of government support is that unemployment figures often did not reveal the true state of affairs, as either displaced women workers were not counted or the steady net outward migration softened the impact. Replacement activities have developed but they have not compensated for the loss of jobs in the old industries (Table 2). In general, the rate of formation of new firms was lower than in other parts of the country, while new technology firms often saw growth siphoned off to other parts of the country. The region suffered from both the general impact of the north-south divide, with the south attracting more of the newer industries, and competition from areas like Merseyside and north-east Wales where high government grants were available for inward investment.

Despite the general economic stagnation of the region there were many significant developments during the twentieth century. Rising incomes, smaller households and easier transport encouraged new house building from the 1920s onwards. Urban sprawl became evident in the interwar period resulting in a coalescing of settlements within a 16km radius of the city centre (Figure 5). By the 1950s a green belt was proposed to control growth. As well as curbing growth on the popular south side and forcing up house prices, this led to the population being diverted to northern areas such as Bolton and Bury. By the 1960s there were plans for motorways in the region: the M62 links Liverpool to Hull, passing through the northern part of the conurbation and incorporating part of an orbital route, the M60 (Figure 5). After 30 years in

© Aidan O'Rourke

Figure 5: Greater Manchester showing built-up area.

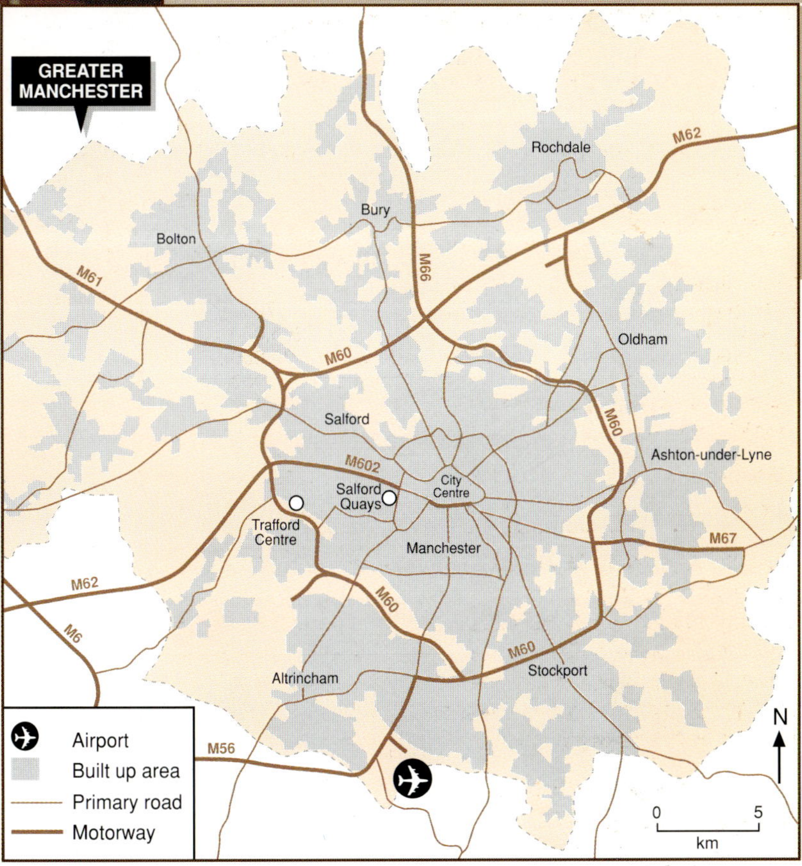

construction it was completed in 2000. Only the M602 penetrates into this ring and access to the city centre from the motorways is relatively poor.

One legacy of the nineteenth century was sub-standard housing, most of which was demolished through slum clearance programmes. These began at the end of the nineteenth century, but did not did not take place on a large scale until the inter-war period. A necessary co-requirement of such programmes was land and new housing for the overspill population. Under the leadership and generosity of Lord Simon, Manchester City Council made plans to build a garden city at Wythenshawe, thirteen kilometres south of the city centre, then in the county of Cheshire. Despite much controversy, work began in the 1930s and was completed in the 1970s when the population peaked at 92,000. In the 1950s another round of slum clearance began, replacing cleared areas with prefabricated, high-rise buildings. These proved to be both badly constructed and unpopular, and by the mid-1990s areas like Hulme had to be cleared once again. The consequences of population decentralisation and slum clearance plus smaller household size resulted in a much reduced population. By the 1970s the inner-city area had a population of about 300,000 compared to the one million at the time of the First World War. As in other cities, hardly anyone lived in the city centre.

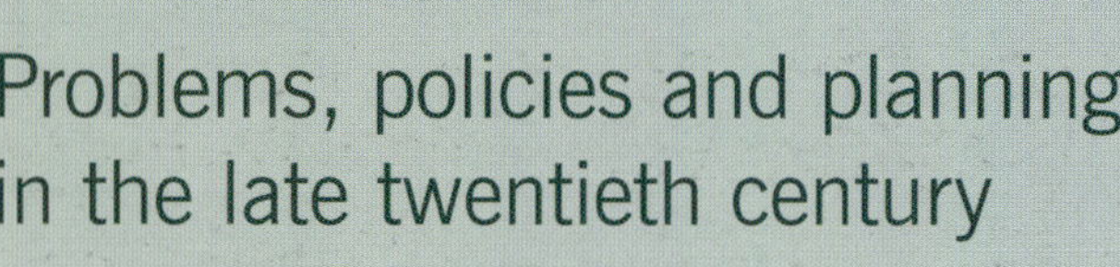

Problems, policies and planning in the late twentieth century

Contents

© University of Manchester

© G-Mex Ltd.

The decline of old industries continued during the late twentieth century, often at an accelerated pace, and particularly during periods of recession such as 1980-82. Cotton textiles, clothing, iron and steel, and heavy engineering were particularly badly affected. The Trafford Park Industrial Estate, which at its peak in the early 1940s had employed 75,000, saw the number of jobs fall to around 25,000 by the mid-1980s. In east Manchester alone over 20,000 jobs were lost in the 1970s and 1980s. City centre employment fell from 160,000 in 1961 to 99,000 in 1991. Major losses occurred in clothing manufacture, newspaper printing, warehousing and transport. In some cases, firms which sought proximity to good road communications, such as newspaper printing, relocated to the suburbs.

In the 1960s planners anticipated retail growth in the city centre based on the assumption that the population would continue to grow, as would the desire to shop in the downtown area. Both these assumptions proved to be wrong. The population in the region was either static or declining: the state of the economy and the continuing process of decentralisation resulted in a reduction in the number of inner-city residents. Retailing instead expanded in the growing outer suburbs, such as Bolton, Stockport and Altrincham. Plans for city centre expansion included clearing the area north of Market Street where, during the 1970s, the 1 million sq ft Arndale Centre was constructed (Figure 6). This attracted trade away from other zones, and roads such as Oldham Street went into decline. City centre office space stayed static, and some older premises remained unoccupied. However there was an expansion of offices in suburban areas, mainly to the south of the city, which were attractive to office developers and firms as they were near the airport, the railway to London, the newly constructed orbital motorway and the homes of many of the executives. For many years rents were also significantly lower. Firms remaining in the city centre included finance, accountancy and law, for whom face-to-face contact is often necessary and a city centre location carries prestige.

Figure 6: City centre: general features.

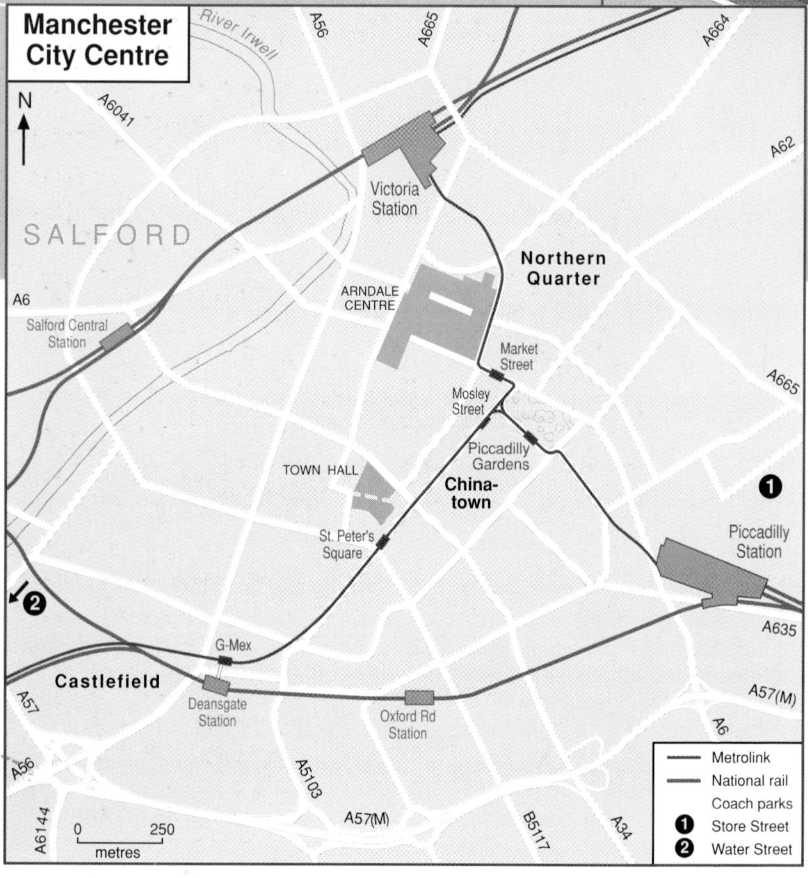

Transport facilities in the city centre were also in decline during this period. Central and Exchange Stations closed in 1969, as did many of the goods depots. By the late 1970s there was evidence of decline on the fringe of the city centre in the form of unused buildings and vacant land. Much of this vacant land became used for car parks, encouraging car commuting which in turn increased traffic congestion.

Planning in the 1970s and early 1980s

In 1974 the new Greater Manchester Council (GMC) was established, with responsibility for strategic functions such as planning and the production of a structure plan. A prime concern for the GMC was to prevent urban sprawl by encouraging development in the core area and by redefining the green belt. It was also concerned about the location of key activities (e.g. shopping and offices) which it considered should be based upon the notion of a hierarchy of centres including the regional centre (Manchester) and the sub-regional centres (such as Bolton and Stockport) as opposed to out-of-town-centre activities such as supermarkets and retail parks.

Inner city policies

The state of Britain's inner cities was an issue of great concern during the 1970s. The focus on issues of poor housing and overcrowding had been replaced by

© University of Manchester

© Arndale Centre

population decline and increasing poverty in inner-city areas. The 1978 Inner Areas Act established partnerships between the government, local authorities and other public bodies to renew these areas through economic, social and environmental policies. An Inner City Partnership area was established in Manchester and Salford covering the district within a radius of about 5km from the city centre. The Conservative government elected in 1979 continued the Inner City Partnerships but placed greater emphasis on economic rather than social policies. Enterprise Zones (EZs) were created to attract economic development through special tax concessions, with other corporate tax advantages to companies investing in new buildings. An EZ covering vacant parts of Trafford Park and the Salford docklands successfully attracted factories, warehouses and offices (see Trail 4, pp. 43-47). However, there were disputes about the extent to which the tax incentives encouraged new economic activities to locate in the region.

The government then introduced Urban Development Corporations (UDCs) to regenerate run-down inner-city areas. In the first round in 1980 only two UDCs were established in the London and Merseyside docklands. However, urban programme grants were available elsewhere and it was under this grant scheme that, in 1985, Salford City Council obtained £30 million to redevelop the Salford docks, renamed Salford Quays. In the mid-1980s the UDC programme was re-established, and Trafford Park Development Corporation was created in 1987 and Central Manchester Development Corporation (actually covering only the east and south of the city centre) in 1988. A further government policy provided 'urban development grants' (later renamed 'city grants') for private developers in inner-city areas to 'lever' investment.

As mentioned above, city centre planning in the 1970s and early 1980s was the responsibility of both the Metropolitan and City Councils. A key feature of the plan was the improvement of transport and accessibility, and in the mid-1980s a light rail system was proposed, to link Victoria Station, the Bury line, Piccadilly Station and the south-west suburban line to Altrincham.

© University of Manchester

Government funding was obtained at the end of the 1980s, and the Metrolink system finally opened in 1992. It proved a great success, doubling the number of passengers carried on the two suburban lines to approximately 14 million passengers a year, and encouraging a large number of car drivers to abandon their vehicles in favour of public transport. Subsequently, a branch to Salford Quays and Eccles was constructed and opened in 1999/2000. Another area where progress was slow was on the ring road around the city centre. As part of a plan to keep 'through' traffic out of the city centre, the first stage of the Mancunian Way had been opened in 1966. However, progress towards completion has been erratic and by 2001 the Salford section has yet to be built.

As part of the plans to regenerate the city centre, the City Council established a City Centre Campaign in 1981, in an effort to win shoppers back, with Christmas shopping campaigns involving lights, decorations and special events. Meanwhile, the GMC concentrated on the re-use of empty or under-used buildings on the fringe of the city centre. In 1979 it bought the Central Station-Great Northern Warehouse site and the Liverpool Road Goods Depot (containing the original 1830 railway passenger station), and in 1982 the Castlefield area was designated as an Urban Heritage Park. In the early 1980s the GMC formed a partnership with Commercial Union to redevelop Central Station as an exhibition centre (G-Mex), and relocated a small science and technology museum to Liverpool Road as part of an ambitious project to create a large Science and Industry Museum.

Planning in the late 1980s and 1990s

The 1980s saw the creation of distinctive government approaches and policies to deal with the problems of city centre and inner city decline, with emphasis on land and property development. The key to success was thought to lie in the involvement of the private sector, either as partners in schemes or as direct investors. The role of the public sector was to prepare sites and provide infrastructure and, if necessary, grants to lever the investment. Although it was hoped that this would increase employment, there was no direct link between levels of investment and the number of jobs created. The link with property development inevitably meant that the outcomes were closely related to market conditions: when the economy was booming, investment would be high, but during a recession developments would stall. Similarly, when the market favoured office buildings, investment would be of this type, and likewise with

© G-Mex Ltd.

leisure, retail and residential investment. Planners therefore had to be responsive to market trends and it was not always possible to set out a vision for an area and achieve the desired outcomes.

Government policies also sought to encourage cities to become 'entrepreneurial'. It was envisaged that cities would actively seek to attract investment and activities, and be competitive in a world where many funds and activities were footloose, and where image, visibility and the lifestyle offered to executives would play a significant role.

In the mid-1980s Manchester could not be described as an entrepreneurial city. Council leaders were concerned with social policies such as equal opportunities which many considered peripheral for an area with high unemployment and much property dereliction. The conversion to a more entrepreneurial focus took place after 1987 and was connected to a number of events. Following the third election victory by the Conservative Party it became clear that oppositional politics achieved little and were against the interests of the local population as far as receiving funds from central government was concerned. Concurrent with these political trends, a local theatre manager, Bob Scott, had almost independently bid for Manchester to host the Olympic Games. The first two bids were not successful, but by the time of the third bid in 1992-93 for the 2000 Olympics, the Council was fully on board and saw it as part of its strategy for the development of the city. With the case of urban regeneration put to an increasingly pro-sport government, Manchester obtained grants to build facilities to support its bid, which enabled the building of an arena and a velodrome. In spite of this effort Manchester still came a poor third to Sydney and Beijing. However, failure did not diminish the city's determination to win a sport mega-event. This was achieved with its bid for the 2002 Commonwealth Games, which, like many other special events, has acted as a catalyst for the improvement of the city centre, using grants from the government and lottery to create a clean and exciting space that will impress visitors.

A third factor in this change in approach was the involvement of leaders on the board of the Central Manchester Development Corporation (CMDC) which was established in 1988. The early success of the CMDC in winning investment revealed the benefits of the entrepreneurial approach, which could be adopted more widely by the City Council. Their reaction to the damage caused by an IRA bomb which exploded on Corporation Street (beside Marks &

© Aidan O'Rourke

© University of Manchester

Spencer) on 15 June 1996 reveals the extent of the Council's conversion. A public-private partnership organisation, Manchester Millennium, was quickly set up to plan for the zone affected and to seek government funds (see Trail 1, pp. 28-33).

Manchester city centre at the Millennium

This section examines first the effect of the contemporary trends of leisure and tourism on the city centre, and second, the residential function of the area. The two are interconnected, with people attracted to living in the city centre because of the leisure amenities available, and leisure amenities stimulated by the presence of a large local population.

Leisure and tourism

In terms of leisure facilities the city centre has always been well supplied – by theatres, cinemas, the Free Trade (concert) Hall, pubs, clubs and restaurants. Their number and range changed little until the 1960s and 70s, when decentralisation of the population led to some decline. This decline was also due to a national rise in the popularity of home leisure activities such as television. The 1980s saw a switch back to more non-home and commercial leisure activities, associated with trends such as increasing affluence, expectations of greater excitement in life, the expansion of the young adult group (18-30) and the rise of clubbing.

In Manchester many of the leisure facilities were located to the south of the city centre, particularly along Oxford Street, although clubs tended to locate in the basements of old warehouses on the east side. By the mid- to late-1980s both the club and music scene in Manchester was thriving, gaining both a national and international reputation. The scale of the industry benefited from the good public transport system of the city centre and the growth of the student population in the late 1980s and early 1990s. By 1998 the four universities and college of music had a combined student population of 83,350 of whom between a half and two-thirds would be living away from home in the inner city. Eating out was also increasingly popular, with the already wide range of restaurants attracting customers from the suburbs and beyond. Another distinctive feature of the leisure scene is the 'Gay Village' around Canal Street. After a period of decline, Manchester's theatres experienced a revival, partly assisted by public policies. Licensing policy was relaxed in the early 1990s to assist the development of a 24 hour city. The expansion of leisure activities is most visible in the growth of pubs and restaurants. With the exception of residences, nearly every new

© Aidan O'Rourke

© Aidan O'Rourke

development appears to allocate some of its space to bars and restaurants. There has also been a revival in the popularity of cinemas. Multiplex cinemas have opened at the Printworks and the Great Northern Experience, reversing the shift to the suburbs.

Since the early 1980s Manchester has made great efforts to develop the tourism industry to boost the city's economy and to create jobs. Conference and exhibition facilities have been created at G-Mex, the new convention centre, and at the universities, including UMIST on the edge of the city centre. Theatre, concerts and sport all attract visitors to the city, many of them staying for a weekend. Hotel room capacity in the city centre has grown from 1400 in 1980 to 3150 in 2000 and will reach 4000 by 2002 if all the plans for new hotels are realised. Special events also attract visitors and the city hopes that the forthcoming Commonwealth Games will put Manchester on the tourist map both in the short and long term.

Living in the city centre

In the mid-1980s the Phoenix Initiative, a private sector body linked to the construction industry, was investigating the possibility of converting warehouses along the Whitworth Street corridor into flats. It concluded that urban development grants would be necessary and put the case to the government. The CMDC later became involved, securing investment for such projects. Thereafter there was a steady trickle of residential developments, either conversions or new buildings. These all proved successful and by the early 1990s the property industry was confident enough to proceed with schemes without the support of subsidies.

In the late 1990s the demand for flats in the city centre took off. Office blocks were converted, often with penthouse flats added, and there was concern that, by forcing up land values, this was harming the office sector. The population of the city centre is estimated to have reached 10,000 by 2000 and could reach 15,000 by 2003. This scale of demand, and the prices being paid, were certainly not anticipated a few years ago. Recent surveys have shown that most of the new residents are young. Men form a clear majority and one survey suggests that one quarter are gay. There are also many 'empty nesters' who may have another home in the countryside. While many of the residents work in the city centre, many commute to work elsewhere, choosing to live in the centre for its lifestyle. There is, however, a potential conflict between the residents of the city and leisure consumers which will require careful planning and control.

Small area studies and trails

Contents

© Len Grant/The Lowry

Trail 1

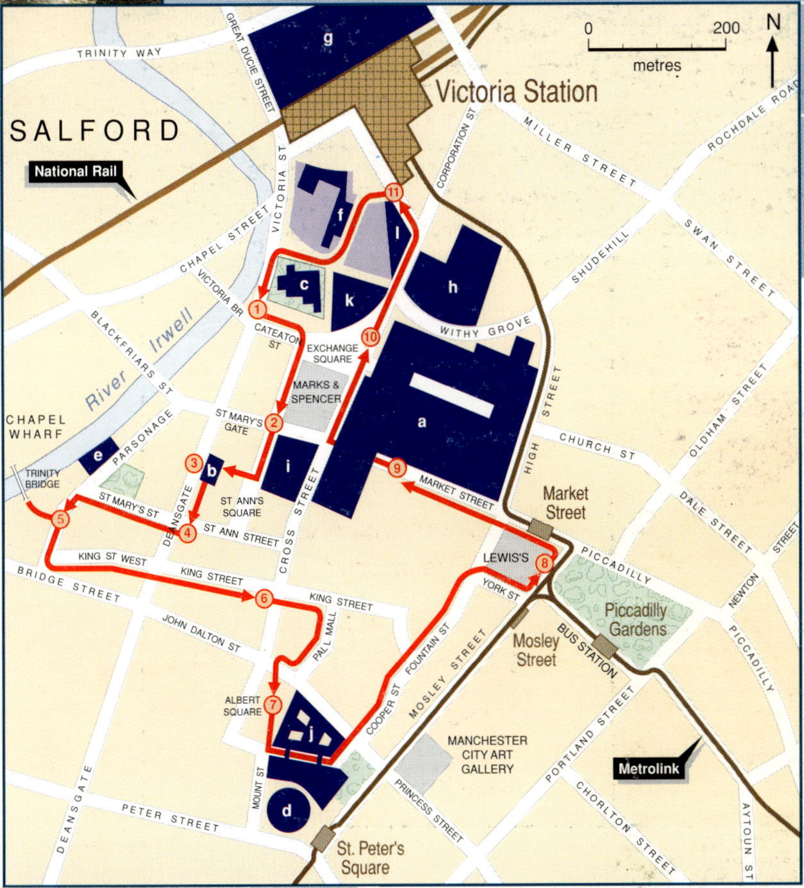

Main Buildings

a. Arndale Shopping Centre
b. Barton Arcade
c. Cathedral
d. Central Reference Library
e. Century House
f. Chetham's School of Music
g. Manchester Indoor Arena
h. Printworks
i. Royal Exchange
j. Town Hall
k. Triangle (former Corn Exchange)
l. Urbis Centre (under construction)

City centres usually contain identifiable specialised zones, and Manchester is no exception. In the centre, an office zone and a retail zone can be identified, the latter divided into the 'popular shopping' area of Market Street and the more up-market shopping area around St Ann's Square, King Street and Deansgate. The outer zones, which were under-used and partially derelict for many years, are now occupied by a combination of leisure and residential quarters. Distinctive zones include Chinatown, the Gay Village, Castlefield, the Northern Quarter and the emerging Convention Quarter. Trails 1-3 below focus on these city centre zones, while Trail 4 focuses on Salford Quays, an area whose function has completely changed.

Town Hall and Library.

Trail 1: The city centre core: Historic centre and planned improvements

Distance: 3-4 kilometres

Walking time (without stops): 45 minutes

Disabled access: yes

Introduction

This trail focuses on the historic core of Manchester and notes past and present changes, and examines the reasons for and implications of such developments.

Retailing and office activities have traditionally provided the prime and central functions of the city centre. Both shops and offices tend to be located in clusters, although the reasons for clustering are different. Shoppers appreciate the grouping of shops within easy walking distance of one another, particularly as it enables them to compare the various offerings, and stores seek to be where there is a high 'footfall'. Within the retail zone there may be further clustering to create sub-zones occupied by shops of a similar kind. In Manchester we can note the separation of popular and up-market shops. Offices may cluster together because of the need for face-to-face contact between representatives of different activities (e.g. lawyers and accountants), or in particular areas for reasons of prestige or visibility.

1. The trail starts at the western end of Cateaton Street overlooking Victoria Bridge. This was the site of the medieval settlement (see Figure 3), and the parish church of St Mary's, which was rebuilt in the fifteenth century as a collegiate church, and

© University of Manchester

© Aidan O'Rourke

raised to the status of Cathedral in 1847. The timber-framed Old Wellington Inn and Sinclair's Oyster Bar were moved to their present sites in the late 1990s as part of Manchester Millennium's project to reshape the city centre following the damage caused by the 1996 IRA bomb and to create a historic quarter (see Chetham's below). On the Salford side of Victoria Bridge is Highland House, a 23-storey office block built in 1968, which after being vacated in 1998 was converted to a hotel on the lower floors and apartments on the upper ones.

2. Turn right onto Cathedral Street a new street linking St Ann's Square to Exchange Square. The area has twice been rebuilt, first following the Second World War and second following the IRA bomb. To the left is the new Marks & Spencer, which opened in late 1999. The largest in Britain, it is hoped it will be a strong anchor, helping the city centre to compete with shopping centres in the surrounding region. There are plans for more shops, including a branch of Harvey Nichols, the famous Knightsbridge store, plus a tower with apartments. The Royal Exchange Building, where cotton was traded, was constructed between 1914-21. Trading ceased in 1972 and it was decided to use the trading floor as a theatre. The Royal Exchange Company began producing plays in the 'theatre in the round' in 1976. The building is open during the daytime and is worth a visit. Also visible from this spot is the Arndale shopping centre (bridging Market Street) the western face of which was extensively remodelled following the bomb damage in 1996.

3. Walk southwards into St Ann's Square, dating back to the beginning of the eighteenth century. Only the church of 1712 survives from this period. In the 1980s the square was pedestrianised and in the 1990s a fountain added. Turn right into Barton Square and enter the Barton Arcade (1871), a good example of a late-nineteenth century enclosed shopping complex. This part of town is noted for its up-market shops, though some in Deansgate have closed down and been replaced by bars and restaurants.

© Aidan O'Rourke

4. Passing into Deansgate, notice Kendal's (formerly Kendal Milne's), Manchester's premier department store, which is now part of the House of Fraser chain. Turn right and follow St Mary's Street to Parsonage Gardens, the site of St Mary's church. Beside the gardens is Arkwright House, built in 1928 as the headquarters for English Sewing Cotton, and now a bar and general offices. To the west of Parsonage Gardens is Century House, formerly the headquarters of an insurance company. Now converted into flats the added penthouse flat was sold in 2000 for £1 million.

5. To the south of Century House is the new Trinity pedestrian bridge over the River Irwell. For decades the Salford side of the river was occupied by railway property, factories and warehouses. When these closed down the land was often left vacant and used only for car parking. In the economic boom of the early 1970s plans were made to build a huge office complex on the Salford side of the river, but the only buildings to materialise were Riverside and Washington Houses, both of which can be seen to the left. Salford City Council has been keen to revitalise this part of the city and in the mid-1990s established a new initiative, the Chapel Street Regeneration Programme, named after the road which runs parallel to the river. A new office complex has been built here for the Inland Revenue, enabling it to house several previously dispersed functions under one roof and in modern conditions. (Note that public offices are often sited on the fringe of the city centre where office land and rents will be lower). A new 5-star hotel, The Lowry, is being built on the riverbank.

6. King Street is part of the up-market shopping area. The central part was pedestrianised in the early 1980s. Until recently, the upper part was the heart of the banking quarter. Local and regional banks were established here in the eighteenth and nineteenth century and a Bank of England branch (at number 82) was built in 1845-46. During the twentieth century these banks were taken over by the national banks who in recent years have restructured their activities and gradually moved away from the

© Aidan O'Rourke

© Aidan O'Rourke

street. Other buildings in the street include insurance offices, the former headquarters of the Ship Canal, and the Reform Club which was opened in 1870 as a meeting place for businessmen. During the late 1990s, this historic heart of the office quarter underwent a remarkable transformation when the lower floors of premises were taken over by designer shops such as Armani which may give the city centre an edge over its suburban competitors.

7. Walk via Pall Mall Street and Clarence Street to Albert Square. This open space was created in the 1860s, then on the edge of the city centre, to be the site for the Town Hall. The Town Hall was designed by Alfred Waterhouse in a French Gothic style and completed in 1877. Its grandeur and pretentiousness are typical of many of the mid-nineteenth century buildings in northern new industrial cities that had gained municipal status and wished to have a landmark building to symbolise this civic pride. Albert Square has changed much since the 1960s; the Town Hall has been cleaned, a pedestrian square created and office blocks built on the warehouse sites opposite the Town Hall. More recently, cafés and restaurants have come to occupy the ground floors.

8. Pass round the side of the Town Hall, along Lloyd Street, next to the extension (1938) and round the back of the building where the Peace Gardens were created in the 1980s. Cross Princess Street and pass along Cooper Street and Fountain Street. Turn right along York Street to Mosley Street and Piccadilly Gardens. This part of the office quarter has been extensively redeveloped; there are several new buildings and many of those that appear old have in fact been redeveloped behind a retained facade.

   Piccadilly Gardens was the site of the Royal Infirmary, built in 1755, which was demolished and relocated in 1908. In recent years this piece of open space, reduced in size by the bus and Metrolink stations, has become valued, but its appearance has left much to be desired. In 1999 the City Council decided that improvements should be made to

© Aidan O'Rourke

© Aidan O'Rourke

create gardens of a high standard which would impress visitors, bearing in mind the impending Commonwealth Games. However, to pay for the restyling it was decided to sell off a small strip on the eastern side for a landmark office building. This created a controversy about the loss of part of the open space and the blocking of the view of the gardens as visitors arrived from the Piccadilly Station. The large buildings to the south, Piccadilly Plaza, were constructed in 1965 and replaced earlier warehouses that were destroyed in the war. The shops on the ground and upper floors have never been very successful in spite of their location near the transport interchange.

9. Turn left along Market Street to the Arndale Centre, through which a detour can be made. The top end of this street is 'anchored' by the department stores of Debenhams and Primark. The design of the Arndale Centre, constructed in the 1970s, is typical of the period, with low ceilings and much less light than later shopping centres (compare The Trafford Centre). It was remodelled in the 1980s to tackle some of these defects, and again after the 1996 bomb.

10. Turn right along Corporation Street to the new Exchange Square. As part of the redesign of the city centre, the pedestrianised Exchange Square has been created, again featuring public art. On the north side is the Royal Exchange (1903) which was badly damaged in 1996. After extensive renovation it was opened in 2000 as The Triangle, a shopping centre for designer clothes. This development, and that of the west Shambles, represent the northward shift of the up-market shopping area. On the north-east corner of Exchange Square is the Printworks, a new entertainment complex opened in 2000 which includes a 20-screen multiplex cinema, an IMAX screen, health and fitness centre, a nightclub, and pubs and restaurants. Once again an old facade has been retained. Before the 1980s this was the largest newspaper printing centre in the north of England where the northern editions of national newspapers were printed. Following changes in printing technology, the

© Aidan O'Rourke

© Aidan O'Rourke

use of labour and the need to distribute the papers from nodal points on the motorway network the works was closed with the loss of 5000 jobs for the city centre.

11. Pass along Corporation Street to Todd Street, Long Millgate and Fennel Street. Buildings associated with the Co-operative movement have long dominated this part of the city. The headquarters of the Co-operative Bank can be seen on Corporation Street and behind it are the tower of the Co-operative Insurance Society and the old offices of the Co-operative Wholesale Society. To the north lies Victoria Station, one of the two main railway stations still operating, albeit downgraded these days. To the west is Chetham's College, one of the oldest buildings in the city, dating from 1422. It houses a library and a specialist school of music. As part of the Manchester Millennium programme the space between these sites was designated partly as open space and partly for a new visitor attraction, Urbis. Due to open late 2001, this will tell the story of cities, their problems and possible solutions, and should be well worth visiting.

Follow Fennel Street to Victoria Street, past the Cathedral back to Cateaton Street.

Trail 2

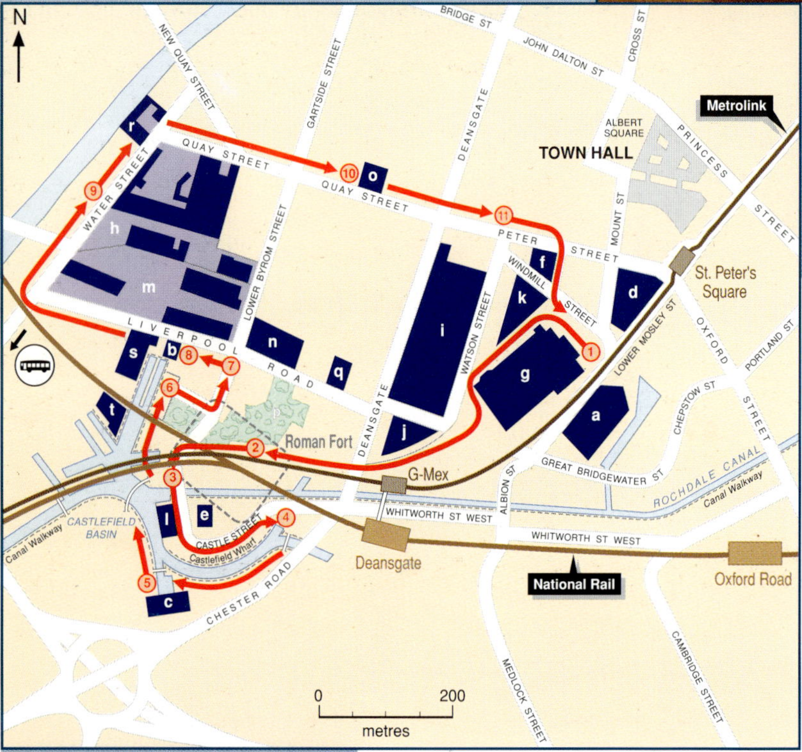

N
BRIDGE ST
JOHN DALTON ST
CROSS ST
NEW QUAY STREET
GARTSIDE STREET
DEANSGATE
Metrolink
ALBERT SQUARE
TOWN HALL
PRINCESS STREET
QUAY STREET
WATER STREET
LOWER BYROM STREET
PETER STREET
MOUNT ST
WINDMILL STREET
WATSON STREET
St. Peter's Square
LIVERPOOL ROAD
LOWER MOSLEY ST
OXFORD STREET
PORTLAND ST
CHEPSTOW ST
Roman Fort
G-Mex
GREAT BRIDGEWATER ST
ROCHDALE CANAL
Canal Walkway
ALBION ST
WHITWORTH ST WEST
CASTLEFIELD BASIN
CASTLE STREET
Castlefield Wharf
Deansgate
National Rail
Oxford Road
CHESTER ROAD
0 200 metres
MEDLOCK STREET
CAMBRIDGE STREET

Main Buildings

a. Bridgewater International Concert Hall
b. Castlefield Visitor's Centre
c. Castle Quay
d. Crowne Plaza Midland Hotel
e. Dukes 92 Public House
f. Free Trade Hall
g. G-Mex Exhibition Centre
h. Granada TV and Studio Tours
i. Great Northern Experience (former Great Northern Warehouse)
j. Hilton Hotel (under construction)
k. International Convention Centre
l. Merchant's Warehouse
m. Museum of Science and Industry (former Lower Campfield Market)
n. Includes Air and Space Museum
o. Opera House
p. Roman Gardens and Wall
q. Upper Campfield Market
r. Victoria and Albert Hotel
s. YMCA Castlefield Hotel
t. Youth Hostel

Trail 2: Castlefield: The renovation of a historic quarter

Distance: 3-4 kilometres

Walking time (without stops): 45 minutes

Disabled access: Possible with small diversions but may be bumpy over cobbled stones

Introduction

Situated in the south-west of the city centre, Castlefield, typical of many of the areas on the fringe of the city centre fell into decay in the 1960s and 70s. Central Station and the Great Northern Warehouse closed in 1969, and the Liverpool Road goods depot in 1975. The Lower Campfield Market (built in 1880) which, as City Hall, had been used for exhibitions was also closed.

© G-Mex Ltd.

Castlefield.

© Jefferson Air Photography, 2000

Older housing was being demolished, many of the canal arms had been filled in and the industry that remained included concrete depots, scrap yards and other unsightly premises. The only 'growing' enterprise was Granada Television which had moved to its site in 1956 and, as well as new buildings, also occupied some old warehouses.

In the 1980s, the area was 'rediscovered' and a number of major changes included:

- In the early 1970s excavations uncovered part of the site of the Roman fort.
- Following the closure of the goods depot a campaign for the preservation of the site.
- In 1978 the GMC purchased both Central Station (including the Great Northern Warehouse) and the Liverpool Road site.
- In 1979 Castlefield was declared a Conservation Area and in 1982 an Urban Heritage Park.
- In 1983 the Museum of Science and Industry was opened in the goods depot and an Air and Space museum in Lower Campfield Market (later united as one institution).
- In 1986 the G-Mex Exhibition Centre was opened in the Central Station.
- In 1988 Granada Studios Tours opened.

Between 1988 and 1996 the CMDC had responsibility for Castlefield and continued to invest in reclamation projects, as well as developing the infrastructure. While much progress has been made, it has been more difficult to develop visitor attractions. In place of these there has been more residential development and also offices and studios. However, the leisure function has grown, as shown by the number of bars and restaurants, and several events are held in the area, such as the Castlefield Carnival. A Castlefield Management Company has been formed to maintain the area and includes a Ranger Service.

© Aidan O'Rourke

© University of Manchester

This trail starts at G-Mex, though an alternative starting place might be the Visitor Centre on Liverpool Road. Note that there is no free parking in the area. Within Castlefield there are several map and information boards.

1. From the front of G-Mex there is a good view of the Bridgewater (concert) Hall opened in 1996, and the 303-bedroom Holiday Inn Crowne Plaza Midland Hotel (built in 1903 as a hotel for the railway station). G-Mex was opened in 1986 as a 100,000 sq ft exhibition centre using the old railway station train shed. A £23 million 800-seat Convention Centre opened in Spring 2001 which, it is hoped, will attract an extra 50,000 conference visitors and in turn boost the city's hotel trade.

   Proceed along the west side of G-Mex to the rear car park and on to the viaduct. This used to carry the trains into Central Station but is now used by the Metrolink route to Altrincham. Look northwards to see the new entertainment complex, the Great Northern Experience. This £100 million development is physically based on the Great Northern Warehouse and contains shops, bars, clubs, restaurants and a car park to which a 24-screen multiplex cinema has been added. A Hilton Hotel is under construction on the site between Great Bridgewater Street and Deansgate.

2. Descend via the stairs or lift. From this point the two parts of the Roman fort which have been reconstructed can be seen. In front of the gateway are the Roman Gardens, laid out in 1983. Proceed under the viaduct to Catalan Square: notice the castle motif on the top of the Cheshire Lines viaduct.

3. During the 1990s this area was greatly transformed. At the entrance to the Rochdale Canal is the pub Dukes 92, the number being the lock number (they were numbered from the Yorkshire side). James Ramsbottom, the entrepreneur behind the pub, also converted the Gail House Eastgate office complex in 1992, and the fire-damaged and

© Aidan O'Rourke

© Aidan O'Rourke

derelict Merchant's Warehouse into offices in 1996. The new sickle span bridge was opened in 1995 and the Barca (bar and restaurant) in 1996. Proceed along Castle Street to Grocers Warehouse. Notice that the old coal wharf is still used for car parking, no more profitable use having been found.

4. Originally, there were several warehouses in this area which were entered from below by short branches of canals (shipping holes), with the goods being off-loaded by pulleys. Only the Merchant's, Middle (Castle Quay) and Grocer's Warehouses survive. The latter has lost its upper storeys but the lower half has been preserved and contains displays to demonstrate how the pulley system worked. Proceed along the southern side of the canal basin (disabled visitors needing a flat pathway should descend via the car park). Notice that the lower part of an old church has been converted into a bar. The £2 million Bass Quay Bar, opened in 1998, can also be seen. On the other side of the Bridgewater Viaduct is Deansgate Quay, opened in 2000 and comprising 102 apartments.

5. Castle Quay (formerly Middle Warehouse) was converted into bars, shops, offices and apartments in 1992. Proceed northwards to Slate Wharf where there are more recent residential units. The development of 102 dwellings, built between 1994 and 1996 involved reclaiming a canal arm. Cross over Merchant's Bridge back to Catalan Square and northwards to the Arena.

6. The canal arms were filled in here and the east side of the site was used for concrete making up to the 1980s. The site has been extensively restored. In 1990 the YMCA moved from its Peter Street location to the Castlefield Hotel. Manchester's first Youth Hostel was opened here in 1994, adjacent to Potato Wharf. Climb the east bank and walk along Duke's Place past the Roman reconstruction to Liverpool Road.

7. Since opening in 1983 the Museum of Science and Industry has had an ongoing programme of development involving the renovation of all the buildings on the site. The museum is well worth a visit. There is also an excellent bookshop on Lower Byron Street, which can be visited without

© Aidan O'Rourke

© Aidan O'Rourke

paying to enter the museum. In 1999, 364,000 visits were made to the museum. Behind the Air and Space section of the museum the residential area of St John's Gardens, built in 1979-81 can be seen.

8. Continue westwards along Liverpool Road. In 2000 a development of 28 apartments, Big M, was completed here. Also note Woolam Place (66 dwellings built between 1993 and 1995), and the entrance to the original railway station. The Visitor Centre has a small exhibition and useful leaflets about the area. Turn right into Water Street.

9. The studios of Granada Television are found adjacent to Water Street and include the sets for Coronation Street, Sherlock Holmes and the Houses of Parliament. After repeated requests for public access, Granada Studios Tours was opened in 1988. Nicknamed 'Hollywood on Irwell' it attracted 700,000 to 800,000 visitors each year for several years. So successful were they that Granada decided to convert some old riverside warehouses into the Meridien Victoria and Albert Hotel in 1992. However, by the end of 1999, visitor numbers had fallen, and the attraction closed for good in 2000.

10. Turn right at the Victoria and Albert Hotel into Quay Street. This zone is undergoing change and renovation. Old office blocks are being refurbished while new ones are being built, sometimes, as with the Skin Hospital, behind old facades. The area to the north of Quay Street is to be redeveloped as part of the Spinningfields scheme and will contain offices and apartments. Continue towards Peter Street, passing the Opera House, built in 1912 on the left.

11. The lower part of Peter Street has been redeveloped as a leisure zone linked to the Great Northern Experience. A new square has been created and there are bars and clubs in this area. The Free Trade Hall, the former concert hall, is being redeveloped as a hotel behind the historic facade. This is all that is left of the original 1856 building, which was constructed on the site of the Peterloo massacre, bombed in the war, then rebuilt in 1952. Return to G-Mex via Windmill Street.

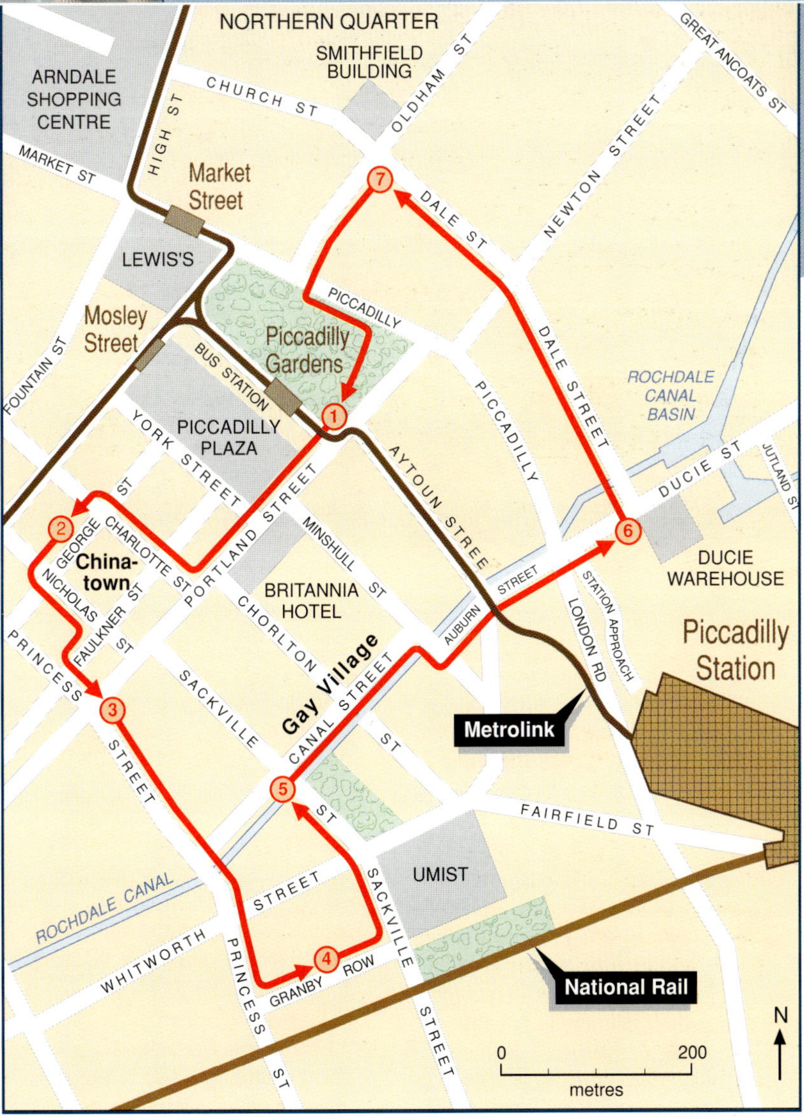

NORTHERN QUARTER
SMITHFIELD BUILDING
ARNDALE SHOPPING CENTRE
CHURCH ST
OLDHAM ST
GREAT ANCOATS ST
MARKET ST
HIGH ST
Market Street
NEWTON STREET
DALE ST
LEWIS'S
PICCADILLY
Mosley Street
Piccadilly Gardens
FOUNTAIN ST
BUS STATION
DALE STREET
ROCHDALE CANAL BASIN
PICCADILLY
PICCADILLY PLAZA
YORK STREET
AYTOUN STREET
DUCIE ST
JUTLAND ST
GEORGE ST
CHARLOTTE ST
PORTLAND STREET
MINSHULL ST
DUCIE WAREHOUSE
China-town
NICHOLAS ST
FAULKNER ST
BRITANNIA HOTEL
CHORLTON ST
AUBURN STREET
STATION APPROACH
LONDON RD
Piccadilly Station
PRINCESS STREET
SACKVILLE ST
Gay Village
CANAL STREET
ST
Metrolink
FAIRFIELD ST
ST
UMIST
ROCHDALE CANAL
STREET
SACKVILLE STREET
WHITWORTH STREET
PRINCESS ST
GRANBY ROW
National Rail
N
0 200 metres

Trail 3: The eastern and northern quarters: The re-use of old buildings

Distance: 3 kilometres

Walking time (without stops): 40 minutes

Disabled access: Yes

Introduction

This trail examines the outer parts of the city centre previously dominated by the warehouses and factories of the cotton and clothing industries. Most of the area was developed or redeveloped in the nineteenth and early twentieth century, but for much of the last century this zone was gradually decaying. Some buildings were demolished and the

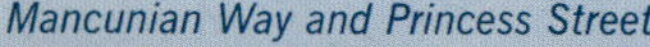

Mancunian Way and Princess Street.

ground used for car parking, and two universities, UMIST and Manchester Metropolitan University, used redundant sites to enable them to expand. Attempts were made to convert warehouses into offices and use old sites for new office blocks, but neither was very successful and many of the old converted warehouses remained underused or empty, though some basements or lower floors were used for clubs and restaurants. For many years the zone's main advantages of low rents and proximity to the city centre core could not be realised due to the poor overall economic climate. In the mid-1980s the situation improved and residential and leisure development has since taken place at an accelerating rate.

1. The trail starts in Piccadilly Gardens. Portland Street was formerly dominated by warehouses, some of which were very grand and faced with stone. On Piccadilly Gardens new offices and hotels have been erected behind the facades. The grandest of all the warehouses was the S & J Watts building of 1851, designed to resemble a Venetian Palace. In 1982 this was converted into the Britannia Hotel. Elsewhere warehouses were demolished in the 1960s and 70s and new office blocks erected.

2. Turn right into Charlotte Street. Notice two examples of fine old warehouses surviving at numbers 14-16 and 10-12. Turn left into George Street to arrive at Chinatown. Chinatown dates from the 1980s and happened spontaneously as the Chinese population in Greater Manchester grew with the opening of Chinese restaurants. On Sundays the Chinese would flock into a Chinese social centre situated in this decayed warehouse zone and from this basis a range of businesses began to develop to serve the community. Later, the City Council saw the opportunity to promote the district and demolished buildings to make a square and to construct an arch. As well as the Chinese Arch, you will see around 20 Chinese restaurants (and an Italian one),

© Aidan O'Rourke

© Aidan O'Rourke

many other Chinese businesses, from food shops to accountants, and also apartments for elderly Chinese.

3. Continue via Nicholas Street and Faulkner Street to Princess Street. This street illustrates the industrial brick architecture of the nineteenth century warehouse zone which, at first sight, seems dull and heavy. However, closer inspection reveals how the flat surfaces are broken up by mixing brick and stone, and various types of decoration. In particular notice the buildings on the north side from 99 to 111. Not until the 1990s did this previously decrepit street start to change. New uses include hotels, flats, clubs and restaurants. Number 103 (former Mechanics Institute) is of particular interest, it being where the first meeting of the Trades Union Congress was held in 1867. Lancaster House on the corner of Princess Street and Whitworth Street is another grand warehouse which has recently been converted into flats.

4. Turning left into Granby Row, Granby House, the first warehouse to be converted into flats (1985) can be seen, the success of which stimulated further conversions in the city centre. While most warehouses can be converted into flats in this way, in some cases demolition and rebuild is more appropriate due to prohibitive cost or the poor state of the building, an example of this can be seen opposite Granby House.

5. Walk along Sackville Street to Canal Street beside the Rochdale Canal and the Gay Village. Until the 1980s the social climate was such that it was difficult for homosexuals to meet in public without fear of discrimination. However, changing attitudes, reinforced by the City Council's Equal Opportunities Policy, enabled pubs and clubs to cater for the gay community without fear of constraint, and as a result, Manchester became famous for being one of the first cities to create a distinctly gay quarter. The reason for its location here is that two public houses, The New Union and The Rembrandt, had a long tradition of catering for the gay community.

6. Follow Canal Street and Auburn Street and cross London Road to reach Ducie Street and Dale Street. In spite of being near the main railway station this area has been

© Aidan O'Rourke

© Aidan O'Rourke

under-used for many years. The old Joseph Hoyle building on Piccadilly remained derelict for years and was an eyesore until the CMDC gave a grant for its conversion into the Malmaison Hotel. The Rochdale Canal basin, which can be viewed from Ducie Street and Dale Street has been used as a car park, but is now to be improved. Look across the basin to the old cotton mills of Ancoats. Many of the old buildings in this area are being converted into apartments, such as the Ducie warehouse which will contain 108 flats.

7. Proceed along Dale Street to Oldham Street and the Northern Quarter. Dale Street and adjacent streets are still associated with the clothing and textile trades, as signs on the buildings show, though clothing manufacture is far less important than it once was. Left behind are many warehouses, including public sales outlets. Oldham Street was once an important shopping district. One consequence of the Arndale Centre, concurrent with the decentralisation of retailing, was the relocation of the main retail stores out of Oldham Street. For 20 years the street has been down-at-heel, with empty shops, 'To Let' signs, cheap discount shops and dowdy bars and restaurants. Since the mid-1990s the City Council has tried to create a distinct character for this area in an attempt to encourage revitalisation. Using the name 'Northern Quarter' was part of this policy, as was promotion of the area as a fashion and music/club zone, the hope being that new businesses would choose to locate here to benefit from clustering. The disappointing results of this policy are visible. On the corner of Church Street and Oldham Street the old Affleck's department store has been converted into flats (the Smithfield building). Further along the street another store has been converted into the Sacha's Hotel. Return to Piccadilly Gardens down Oldham Street.

© Aidan O'Rourke

Trail 4

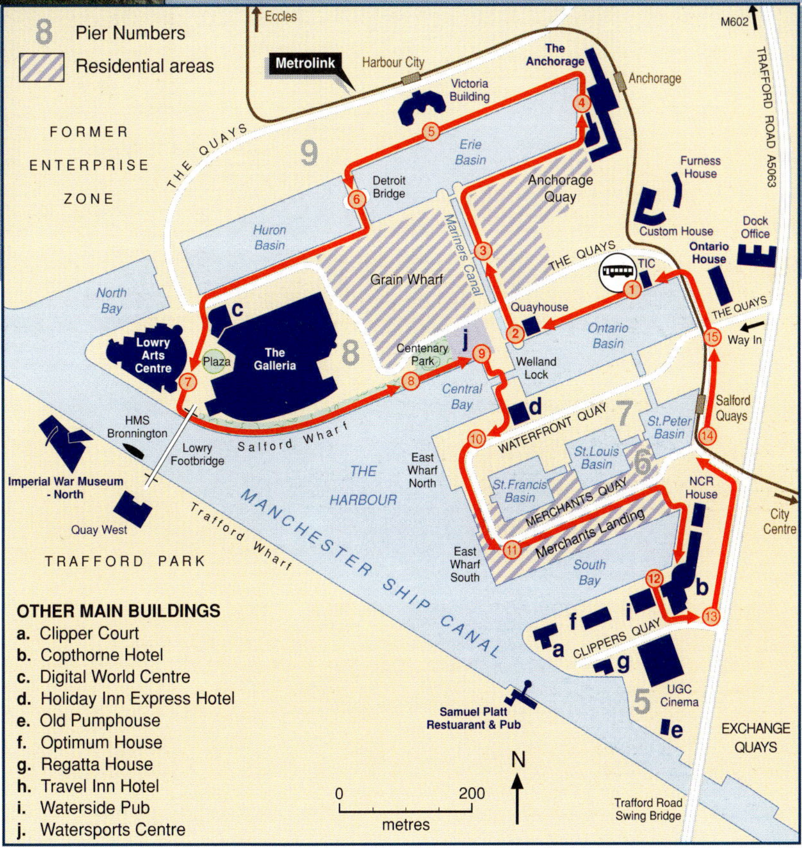

8 Pier Numbers

Residential areas

OTHER MAIN BUILDINGS

- **a.** Clipper Court
- **b.** Copthorne Hotel
- **c.** Digital World Centre
- **d.** Holiday Inn Express Hotel
- **e.** Old Pumphouse
- **f.** Optimum House
- **g.** Regatta House
- **h.** Travel Inn Hotel
- **i.** Waterside Pub
- **j.** Watersports Centre

Trail 4 Salford Quays: The redevelopment of redundant docks

Distance: 3-4 kilometres

Walking time (without stops): 45 minutes.

Disabled access: Yes

Talk and guide available from Salford Quays Tourist Information Centre at a cost and advanced booking.

Introduction

The Manchester Ship Canal was opened in 1894. The main docks were built beside Trafford Road in Salford, and later, in 1905, the large Dock 9 was opened. To encourage trade, an industrial estate (the world's first) was planned in 1896 at Trafford Park on the south bank, with a private railway link to the docks. The re-named basins reflect the strong links with North America. For many decades the docks handled approximately 2 million tons of cargo a year and at their peak employed about 3500. However, in the 1960s and 70s, largely as a result of containerisation, for which Manchester was less suited than other British ports, trade at the docks was much reduced, and between 1972 and 1982 almost completely disappeared.

In 1981 parts of the dock estate, particularly the undeveloped northern

© Len Grant/The Lowry

Salford Quays.

© Jefferson Air Photography, 2000

areas, were designated an Enterprise Zone (EZ) for a period of 10 years. It was clear that trade in the docks was likely to disappear completely and that large funds would be required to transform them for other uses. The Manchester Ship Canal Company, thinking that there would not be a return on any investment, sold the docks to Salford City Council so that public funding could be obtained. The area was then renamed Salford Quays. In 1985 the government agreed to provide £30 million for basic infrastructure, and other funding came from the City Council and the European Union. A condition of this funding was that all development should be privately financed, and to that end the southern part of the docks was sold to Urban Waterside. Unlike many other dockland redevelopment schemes in Britain, no buildings have been preserved except the Dock Offices.

Redevelopment gathered momentum in the late 1980s, and the economic boom and forthcoming end of EZ status in 1991 stimulated construction, particularly of offices. However, the country was facing recession when the offices were just being completed, with the result that many remained unoccupied for some years, and development overall was stalled. In the late 1990s the economy picked up, and development resumed. Lottery funding enabled the construction of The Lowry Arts Centre. Many new dwellings have been built, and the Metrolink extension (opened in 1999) is likely to encourage further development. Office take-up has increased, reducing vacancies and encouraging developers to contemplate new construction.

There are serious concerns about the extent to which the scheme has benefited and will benefit local people,

© Aidan O'Rourke

© Len Grant/The Lowry

as the main beneficiaries of the jobs, housing and leisure facilities have been middle-class people from outside the area. Another aspect of the lack of integration is reflected in the need for a high level of security.

1. The Trail starts at the car park next to the Tourist Information Centre (TIC), which sells leaflets, maps and postcards. There are maps of Salford Quays at various points along the route.

2. Walk along the side of Ontario Basin past the Travel Inn (1996) and the Quayhouse/Beefeater restaurant (1990), both owned by the Whitbread company. Turn right to walk along Mariner Canal.

3. The Grain Wharf estate of houses and apartments was built between 1985 and 1995. Salford Quays tends to attract single people or couples rather than families. Many properties are rented, perhaps as an interim measure, and some belong to people who only stay in Manchester to work for a few days in the week and may have another house elsewhere. In terms of property prices and social composition, this area contrasts sharply with the Ordsall district of Salford, a typical inner-city district with much public housing and lower-income residents, on the other side of the Trafford Road.

4. On reaching Erie Basin turn right past Anchorage Quay, a recent development of 80 dwellings completed in 1998 and proceed to the large office block at the head of the dock, called the Anchorage. In 1991 the Anchorage was the first large office block to be constructed in the Quays. Its main tenant is BUPA, with over 500 employees.

5. Pier 9 on the north side of the Erie Basin is still owned by the Manchester Ship Canal Company. In the late 1980s the company planned Harbour City, a large office complex along the entire length of the quay. The 116,000 sq ft Victoria Building was completed in 1992 in the middle of a recession, and the rest of the scheme is awaiting an upturn in the office market. In mid-2000 the company produced new proposals for houses, shops and offices along Pier 9 and the Ship Canal.

6. Turn left to cross Dock 9 over Detroit Bridge, a former railway bridge that was relocated here from a site near Clippers Quay. Dock 9 is the largest and can be used for events such as regattas.

7. Continue along Huron Basin to The Lowry Arts Centre. Designed by Michael Wilford and opened in April 2000, the £98 million centre (£64 million of which was Lottery funded)

© Aidan O'Rourke

© Aidan O'Rourke

contains two theatres, the Lowry art collection and other exhibition space. A visit is recommended. Next to The Lowry, the Galleria is a £70 million complex containing cinemas, shops, restaurants bars and an apartment block, opened in 2001. A third development at this end of the pier is the Digital World Centre.

8. Take the footpath past the Lowry footbridge to Centenary Park. Across the bridge the Quay West office building can be seen, as can HMS Bronnington (Prince Charles' old ship) and the striking Imperial War Museum of the North designed by Daniel Libeskind in the form of shards (broken pieces of a globe) completed in 2001. Behind it is the Hovis flour mill built in 1907, and the only survivor of Trafford Park's three mills. Today the mill uses imported wheat brought by road from Liverpool. Centenary Park and Walk were constructed in 1994 to mark the anniversary of the canal. Across the canal the Manchester United 'Theatre of Dreams' stadium can be seen.

9. Next to Centenary Park is the Watersports Centre, constructed in 2001 for use by local people. Docks 7, 8 and 9 were dammed to keep out the polluted water from the River Irwell and the Ship Canal, and the water within the basins is purified. From Dock 8 (Ontario Basin) there is an entry lock and the other two docks are linked to the system by newly constructed canals.

10. Cross Central Bay to Pier 7, which is mainly given over to the 'Waterfront 2000' office development. Each building is relatively small and low-rise and many are occupied by computer companies who regarded the Quays as a prestige location. The last site on the pier has been occupied since 1998 by the Holiday Inn Express Hotel.

11. Pier 6 contains one of the two main residential areas, both of which were built on non-EZ land. Merchant's Landing, containing 153 dwellings, was built between 1987 and 1991. More recently houses have been built on the piers in Dock 7. The

© Aidan O'Rourke

type of residents here are as those of Grain Wharf. There is a fairly high turnover of residents in these areas, as shown by the number of For Sale and To Let signs.

12. Walk around South Bay to Pier 5 and the Waterside Pub (1988). This was the first part of the docks to be developed by Urban Waterside. The 8-screen UGC cinema was opened in 1986, followed by the 166-room Copthorne Hotel in 1987. The rest is given over to offices; Clipper Court and Regatta House (1988) and Optimum House (1989), and also Banks pub in the old pumphouse (1992).

13. Turn left from Clippers Quay onto Trafford Road. On the other side of Trafford Road is the large Exchange Quay (530,000 sq ft), one of three such large office buildings constructed in the early 1990s, which contrast with the earlier, smaller-scale office blocks.

14. Turn left onto St Peter Quay. The island site between here and Trafford Road is owned by Urban Waterside. Originally there were plans for a large development here, but the early 1990s recession prevented such proposals from being realised and in its place four low-density eateries were opened in 1996.

15. Continue along St Peter Quay, where The Dock Office can be seen. Built in 1926, part of the Office is still used by the Ship Canal Company, the rest being let. Behind it is Furness House, built in 1969 for use by Furness Withy who ran Manchester Liners, then an important shipping line associated with the port. In front of the Dock Office is Ontario House, built in 1990 and let to a government department. The value of such offices has increased since Trafford Road was converted to a dual carriageway in 1997 and the Metrolink line was opened in 1999, thus enhancing their accessibility. Return to the TIC round Ontario Basin.

Bibliography and further information

© Aidan O'Rourke

Briggs, A. (1968) *Victorian Cities*. London: Penguin.

Canniffe, E. and Jefferies, T. (1998) *Manchester Architectural Guide.* Manchester: Manchester School of Architecture.

Carter, C.F. (1962) *Manchester and its Region.* Manchester: Manchester University Press.

Deas, I., Peck, J., Tickell, A., Ward, K. and Bradford, M. (1999) 'Rescripting urban regeneration the Mancunian way', in Imrie, R and Thomas, H. (eds) *British Urban Policy: An Evaluation of Urban Development Corporations.* London: Sage.

Dennis, R. (1984) *English Industrial Cities in the Nineteenth Century: A Social Geography*. Cambridge: Cambridge University Press.

Engels, F. (1845) *The Condition of the Working Classes in England.* London: Penguin.

Frangopulo, N.J. (1962) *Rich Inheritance: A Guide to the History of Manchester.* Manchester: Manchester Education Committee.

Freeman, T.W., Rodgers, H.B. and Kinvig, R.H. (1966) *Lancashire, Cheshire and the Isle of Man.* London: Nelson.

Gaskell, E. (1848) *Mary Barton.* London: Penguin.

Gaskell, E. (1855) *North and South.* London: Penguin.

Hall, P. (1998) *Cities in Civilisation: Culture, Innovation and Urban Order.* London: Weidenfield and Nicolson.

Haslam, D. (1999) *Manchester England: The Story of the Pop Cult City.* London: Fourth Estate.

Kitchen, T. (1997) *People, Politics, Policies, and Plans: The City Planning Process in Contemporary Britain.* London: Paul Chapman Publishing.

Law, C.M. *et al.* (1988) *The Uncertain Future of the Urban Core.* London: Routledge.

Law, C.M. (1992) 'Property-led urban regeneration in inner Manchester', in Healey, P. *et al. Rebuilding the City.* London: E&F Spon.

McNeil, R. and George, A.D. (1997) *The Heritage Atlas 3: Warehouse Album.* Manchester: Field Archaeology Centre, University of Manchester.

Mitchell, B.R. (1988) *British Historical Statistics.* Cambridge: Cambridge University Press.

O'Connor, J. and Wynne, D. (eds) (1996) *From the Margin to the Centre: Cultural Production and Consumption in the Post-Industrial City.* Aldershot: Arena.

Parkinson-Bailey, J.J. (2000) *Manchester: An Architectural History. Manchester:* Manchester University Press.

© Aidan O'Rourke

© University of Manchester

© John Peters

Rodgers, B. (1986) 'Manchester: metropolitan planning by collaboration and consent: or civic hope frustrated', in Gordon, G. (ed) *Regional Cities in the UK 1890-1980.* London: Harper and Row.

Schofield, J. (2000) *City Life Guide to Manchester.* Manchester: City Life Publications.

Taylor, I., Evans, K. and Fraser, P. (1996) *A Tale of Two Cities: A Study of Manchester and Sheffield.* London: Routledge.

White, H.P. (ed) (1980) *The Continuing Conurbation: Change and Development in Greater Manchester.* Farnborough: Gower.

Williams, G. (1999) 'Greater Manchester', in Roberts, P., Thomas, K. and Williams, G. (eds) *Metropolitan Planning in Britain: A Comparative Analysis.* London: Jessica Kingsley.

Further information

Tourist Information Centres

Manchester Visitor Centre, Town Hall Extension, Lloyd Street, Manchester. Tel: 0161 234 3157 (where a video of Manchester can be seen).

Castlefield, 101 Liverpool Road, Manchester. Tel: 0161 834 4026.

Salford Quays, 1 The Quays, Salford M5 2SQ, Tel: 0161 872 7463.

Maps

Ordnance Survey *Landranger 109 Manchester and Surrounding Area* (1:50,000).

Andrew Taylor *Manchester City Centre* (Principal land uses), Second Edition 1998 (1:3500).

Cityscape *Manchester*, Third Edition 1996 (9.5 inches:1 mile).

Websites

Eye Witness in Manchester www.aidan.co.uk

Manchester Airport: www.manairport.co.uk

Manchester City Council: www.manchester.gov.uk

Manchester City Planning: www.manchesterupdate.org.uk

Manchester Evening News: www.manchesteronline.co.uk

Northern Quarter: www.nqn.org.uk

Englands North West: www.englandsnorthwest.com

Salford City Council: www.salford.gov.uk

Greater Manchester Research Unit: www.gmresearch.u-net.com

BBC Manchester: www.bbc.co.uk/manchester

Commonwealth Games 2002: www.commonwealthgames.com

Local Studies Libraries

Manchester Central Reference Library, St Peter's Square, Manchester

Salford Local Studies Centre, The Crescent, Salford

Trafford Local Studies Centre, Sale

Photo credits

The GA would like to thank the following for their help with photographs for this book:

- Jefferson Air Photography
- Len Grant/The Lowry
- G-Mex Ltd.
- Arndale Centre
- John Peters
- Gary Hanna

With special thanks to Jane Naylor, International & Public Relations Office, University of Manchester, and Aidan O'Rourke, author of *Eyewitness in Manchester.* A personal online view of Manchester with photographs and articles, *Eyewitness in Manchester* joined Manchester Online (the Manchester Evening New/Guardian Media Group) in 1998 and has grown into one of the largest sources of pictures and information on Manchester on the Internet. Visit www.aidan.co.uk